Physical Properties of Diamond and Sapphire

T0138887

Physical Properties of Diamond and Sapphire

Roshan L. Aggarwal
Anant K. Ramdas

CRC Press
Taylor & Francis Group
Boca Raton London New York

CRC Press is an imprint of the
Taylor & Francis Group, an **informa** business

CRC Press
Taylor & Francis Group
6000 Broken Sound Parkway NW, Suite 300
Boca Raton, FL 33487-2742

First issued in paperback 2021

© 2019 by Taylor & Francis Group, LLC
CRC Press is an imprint of Taylor & Francis Group, an Informa business

No claim to original U.S. Government works

ISBN-13: 978-0-367-23508-6 (hbk)
ISBN-13: 978-1-03-217820-2 (pbk)
DOI: 10.1201/9780429283260

This book contains information obtained from authentic and highly regarded sources. Reasonable efforts have been made to publish reliable data and information, but the author and publisher cannot assume responsibility for the validity of all materials or the consequences of their use. The authors and publishers have attempted to trace the copyright holders of all material reproduced in this publication and apologize to copyright holders if permission to publish in this form has not been obtained. If any copyright material has not been acknowledged, please write and let us know so we may rectify in any future reprint.

Except as permitted under U.S. Copyright Law, no part of this book may be reprinted, reproduced, transmitted, or utilized in any form by any electronic, mechanical, or other means, now known or hereafter invented, including photocopying, microfilming, and recording, or in any information storage or retrieval system, without written permission from the publishers.

For permission to photocopy or use material electronically from this work, please access www.copyright.com (http://www.copyright.com/) or contact the Copyright Clearance Center, Inc. (CCC), 222 Rosewood Drive, Danvers, MA 01923, 978-750-8400. CCC is a not-for-profit organization that provides licenses and registration for a variety of users. For organizations that have been granted a photocopy license by the CCC, a separate system of payment has been arranged.

Trademark Notice: Product or corporate names may be trademarks or registered trademarks, and are used only for identification and explanation without intent to infringe.

Publisher's Note

The publisher has gone to great lengths to ensure the quality of this reprint but points out that some imperfections in the original copies may be apparent.

Library of Congress Cataloging-in-Publication Data

Names: Aggarwal, R. L. (Roshan Lal), 1937- author. | Ramdas, Anant K., author.
Title: Physical properties of diamond and sapphire / Roshan L. Aggarwal and Anant K. Ramdas.
Description: First edition. | Boca Raton, FL : CRC Press/Taylor & Francis Group, 2019. | Includes bibliographical references and index.
Identifiers: LCCN 2019006297| ISBN 9780367235086 (hardback : acid-free paper) | ISBN 9780429283260 (ebook)
Subjects: LCSH: Diamonds. | Sapphires.
Classification: LCC QE393 .A384 2019 | DDC 549/.27--dc23
LC record available at https://lccn.loc.gov/2019006297

**Visit the Taylor & Francis Web site at
http://www.taylorandfrancis.com**

**and the CRC Press Web site at
http://www.crcpress.com**

This monograph is dedicated to our parents (Chet Ram Aggarwal and Lila Vati Aggarwal and Lakshminarayanapuram Ananthakrishnalyer Ramdas and Kalyani Ramdas), our spouses (Pushap Lata Aggarwal and Vasanthalakshmi Ramdas), our children (Rajesh Aggarwal and Achal Aggarwal), and our grandchildren (Isha Aggarwal, Neena Aggarwal, Akash Aggarwal, and Ashok Aggarwal).

Contents

Preface

THIS MONOGRAPH IS INTENDED to provide readers with the physical properties of diamond and sapphire, which are both gemstones and have similar properties. Diamond is an optically isotropic crystal. Sapphire is a uniaxial crystal with refractive indices n_E and n_O for polarization parallel and perpendicular to the optic axis, respectively. This monograph includes crystal structure and growth, mechanical properties, thermal properties, optical properties, light scattering, and sapphire lasers. Mechanical properties include hardness, tensile strength, compressive strength, and Young's modulus. Thermal properties include thermal expansion, specific heat, and thermal conductivity. Optical properties include transmission, refractive index, and absorption. Light scattering includes Raman scattering and Brillouin scattering. Sapphire lasers include chromium-doped and titanium-doped lasers. The sources for the material in this monograph are journal articles and others as acknowledged in the references.

Acknowledgments

W E THANK DR. MARC BERNSTEIN for his approval to write this monograph. We thank Dr. William Herzog and Dr. Mordechai Rothschild for discussions regarding this work. We thank Casey Reed for drafting some of the figures.

Authors

Roshan L. Aggarwal retired from Massachusetts Institute of Technology (MIT) effective April 1, 2016, after 51 years of service. He is currently working as part-time flexible technical staff in Group 81 "Chemical, Microsystem, and Nanoscale Technologies" at MIT Lincoln Laboratory. Previously, he was technical staff at MIT Lincoln Laboratory for 30 years (1986–2016), senior research scientist in the MIT Physics Department for 12 years (1975–1987), associate director at the MIT Francis Bitter National Magnet Laboratory for 7 years (1977–1984), and technical staff at the MIT Francis Bitter National Magnet Laboratory for 12 years (1965–1977). Dr. Aggarwal is a fellow of the American Physical Society, a senior member of the Optical Society of America, a fellow of the Punjab Academy of Sciences, and a recipient of the Albert Nelson Marquis Lifetime Achievement Award.

Anant K. Ramdas retired as the Lark-Horovitz Distinguished Professor Emeritus of Physics in 2016 after 60 years of service at Purdue University. He received his PhD in Physics from Poona University, India, in 1956; his thesis advisor was Prof. C. V. Raman. Professor Ramdas is a recipient of the Alexander von Humboldt Foundation Senior U. S. Scientist Award, recipient of the Raman Centenary Medal of the Indian Academy of Sciences, Frank Isakson Prize of the American Physical Society, Sigma Xi Faculty Research Award, Ruth and Joel Spira Award for excellence in undergraduate teaching, and Herbert Newby McCoy Award of Purdue University. Prof. Ramdas is a fellow of the American Physical Society, fellow of the Optical Society of America, fellow of the American Vacuum Society, fellow of the American Association for the Advancement of Science, and fellow of the Indian Academy of Sciences.

CHAPTER 1

Introduction

D IAMOND HAS FASCINATED MANKIND since the early ages for its aesthetic beauty. Diamond is a girl's best friend because it is both expensive and associated with romantic love. Diamond is the birthstone for April.

Diamonds are gemstones, which are crystals of carbon. Diamond crystals are also grown in the laboratory. Natural diamonds come in a variety of colors, including white, yellow (Sobolev and Lavrent'Ev 1971), pink (King et al. 2002), and blue (King et al. 1998). Most diamonds are white. Yellow, pink, purple, and blue diamonds are rare and hence expensive.

Figure 1.1 shows the purplish-red, red, and orangy-red colors of diamonds produced by Lucent Diamonds using a multistep process that involves high pressure–high temperature (HPHT) annealing, irradiation, and low-pressure annealing at relatively low temperatures (Wang et al. 2005).

The wavelength ranges for the red, orange, yellow, green, blue, and violet colors are approximately 675–741, 600–675, 538–600, 482–538, 450–482, and 379–422 nm, respectively (Loeffler and Burns 1976).

Both ruby and sapphire are corundum (α-aluminum oxide) crystals. The birthstone months for ruby and sapphire are July

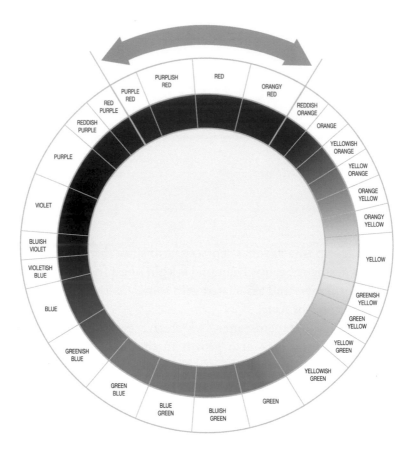

FIGURE 1.1 Colors of diamonds (shown on the color wheel) produced by Lucent Diamonds using a multistep process for natural type Ia diamonds. (After Wang, W. et al., *Gems Gemol.* 41, 6, 2005.)

and September, respectively. Ruby is red (~675–741 nm), whereas sapphire is typically blue (~450–482 nm) but also occurs in yellow (~538–600 nm) and green (~482–538 nm) colors. The red (~675–741 nm) color of ruby is due to chromium. Yellow sapphires are due to iron. Blue and green sapphires are due to iron and titanium pairs. The physical properties of ruby and sapphire are the same.

1.1 DIAMOND

1.1.1 Color of Natural Diamonds

Colors of diamonds are caused by optical absorptions due to impurities (nitrogen, hydrogen, boron) and defects (Fritsch 1998). Nitrogen is responsible for the yellow (~538–600 nm) color of diamonds as shown in Figure 1.2 (Chrenko et al. 1971).

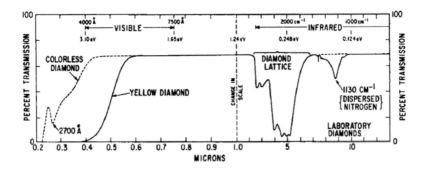

FIGURE 1.2   Absorption spectrum of yellow diamond. (After Chrenko, R. M. et al., *Philos. Mag.* 23, 313, 1971.)

Boron, which shows absorption starting at 3.35 μm and continuing with decreasing intensity into the visible range, is responsible for the blue (~450–482 nm) color of diamonds, as shown in Figure 1.3 (King et al. 1998).

Figure 1.4 shows the ultraviolet-visible-near infrared (UV-V-NIR) absorption spectrum of purple diamonds (Titkov et al. 2008). The absorption spectrum in Figure 1.4 exhibits increasing absorption below 450 nm and a broad absorption band centered at about 550 nm. The N3-related zero phonon line (ZPL) is observed at 415 nm.

1.1.2 Physical Properties of Diamond

Diamond is a large-gap semiconductor with an indirect gap of 5.48 eV (226 nm) and direct gap of 7.12 eV (174 nm) (Logothetidis et al. 1992). Diamond is optically transparent in the ultraviolet, visible, infrared, and far infrared. Diamond has a high refractive

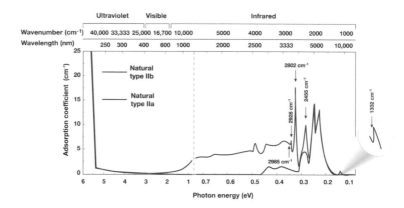

FIGURE 1.3   Absorption spectrum of boron-containing natural type IIb diamond compared with that of natural type IIa diamond. (After King, J. M. et al., *Gems Gemol.* 246, 1998.)

UV-Vis-NIR Absorption Spectrum

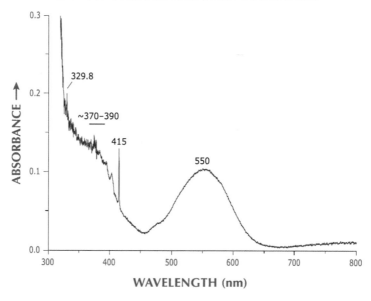

FIGURE 1.4   UV-VIS-NIR absorption spectrum of purple diamonds. (After Titkov, S. V. et al., *Gems Gemol.* 44, 56, 2008.)

index of 2.46–2.41 for visible light (0.40–0.65 μm), which gives cut diamonds their brilliance (Palik 1985).

Natural diamonds, which are formed deep within the earth under extreme heat and pressure, contain unique birthmarks that are either internal (inclusions) or external (blemishes). Diamonds without these birthmarks are rare and very expensive. Diamonds are assigned a clarity grade using the GIA International Grading System under 10× magnification. The GIA covers 12 clarity grades: flawless (FL), internally flawless (IFL), very very slightly included (VVS$_1$ and VVS$_2$), very slightly included (VS$_1$ and VS$_2$), slightly included (SI$_1$ and SI$_2$), and included (I$_1$, I$_2$, and I$_3$).

The density of diamond is 3.515 g/cm^3; that is, it has a C^{12} concentration of 1.762×10^{23} cm^{-3}. The weight of diamonds is expressed in carats. One carat is equal to 200 mg and has a volume of 56.9 mm^3.

Diamond is harder and conducts heat better than any other known material. The relative hardness of diamond is 10 Mohs; Mohs chose 10 well-known minerals and arranged them in order of their scratch hardness in descending order in 1822. The thermal conductivity of diamond at 300 K is 900 W/m·K compared to that of 386 W/m·K for copper (Touloukian et al. 1971). The physical properties of hardness and thermal conductivity make diamond an excellent choice for many technological applications, such as cutting, polishing, protective coatings, and heat conduction. Diamonds used for technological applications are produced using the chemical vapor deposition (CVD) method.

There are four types of natural diamonds: (1) type Ia, which contains pairs and other aggregates of substitutional nitrogen atoms; (2) type Ib, which contains single/isolated substitutional nitrogen atoms; (3) type IIa, which is relatively free of nitrogen and other impurities; and (4) type IIb, which contains substitutional boron atoms.

Most diamonds are excellent insulators. The electrical resistivity is on the order of 10^{11}–10^{18} Ωm. Boron-doped natural blue diamonds are p-type semiconductors. However, certain blue-gray diamonds that contain hydrogen are not semiconductors. Phosphorus-doped diamond films, produced by CVD, are n-type semiconductors.

1.2 SAPPHIRE

1.2.1 Colors of Sapphire

Ruby is red sapphire due to Cr^{3+} ions. Figure 1.5 shows the optical absorption of ruby in the visible range. There are two absorption bands in the violet, and green and yellow regions. There is negligible absorption in the red, which is responsible for the red color of ruby.

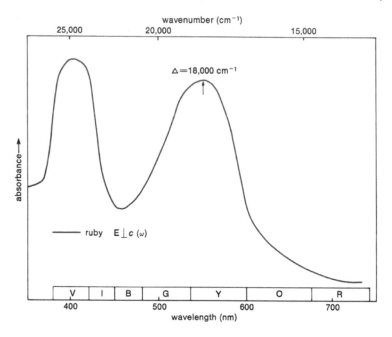

FIGURE 1.5   Absorption spectrum of ruby. (After Loeffler, B. M., and R. G. Burns, *Am. Sci.* 64, 636, 1976.)

Blue sapphires are due to the absorption of Ti^{4+}-O^{2-}-Fe^{2+} at 18,000 cm^{-1} (556 nm) for perpendicular-to-c polarization and 14,200 cm^{-1} (704 nm) for parallel-to-c polarization. Green sapphires are due to the absorption of Fe^{2+}-O^{2-}-Fe^{3+} at 11,500 cm^{-1} (870 nm) for perpendicular-to-c polarization and 10,000 cm^{-1} (1000 nm) for parallel-to-c polarization. Figure 1.6 shows the room-temperature absorption spectrum of blue-green sapphire for perpendicular-to-c polarization (Ferguson and Fielding 1971).

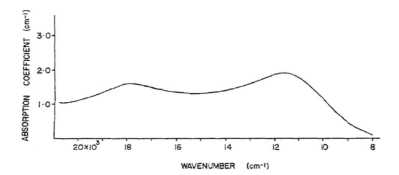

FIGURE 1.6   Room-temperature absorption spectrum of a natural blue-green sapphire for the perpendicular-to-c polarization. (After Ferguson, J., and P. E. Fielding, *Australian J. Chem.* 25, 1371, 1972.)

Yellow sapphires are due to the absorption of single Fe^{3+} ions and pairs of Fe^{3+} ions (Ferguson and Fielding 1972). Figure 1.7 shows the room-temperature spectra of yellow sapphire containing 0.99% Fe for (a) parallel-to-c polarization and (b) perpendicular-to-c polarization.

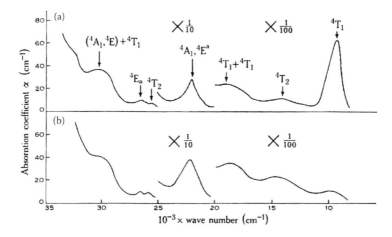

FIGURE 1.7   Absorption spectra of a natural yellow sapphire containing 0.99% Fe by weight. (a) Parallel-to-c polarization, and (b) perpendicular-to-c polarization. (After Ferguson, J., and P. E. Fielding, *Australian J. Chem.* 25, 1371, 1972.)

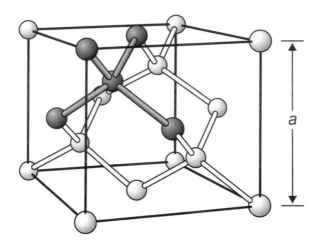

FIGURE 2.1 Crystal structure of diamond showing the tetrahedral arrangement of carbon atoms.

(a)   (b)

FIGURE 2.2 Artificial diamonds: (a) 1-mm diamond shown with phonograph needle, and (b) 0.2–0.5 mm octahedra. (After Bundy, F. P. et al., *Nature* 176, 51, 1955.)

Figure 2.3 shows the x-ray diffraction pattern of artificial and natural diamonds.

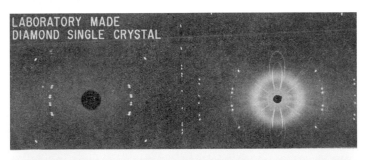

FIGURE 2.3   X-ray diffraction pattern of laboratory-made (artificial) diamond and natural diamond. (After Bundy, F. P. et al., *Nature* 176, 51, 1955.)

Chang et al. reported diamond crystal growth by plasma chemical vapor deposition in 1988 (Chang et al. 1988). Figure 2.4 shows a schematic of the 2450-MHz discharge tube reactor used by Chang et al.

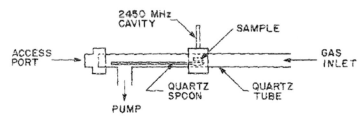

FIGURE 2.4   Schematic of the 2450-MHz discharge tube reactor. (After Chang, C.-P. et al., *J. Appl. Phys.* 63, 1744, 1988.)

A single-crystal diamond growth rate >20 mm/hr was achieved using this reactor. Figure 2.5 shows a high-quality single-crystal diamond grown using 100 H_2/4 CH_4/0.5 O_2 sccm at 30 Torr with a 100-W discharge.

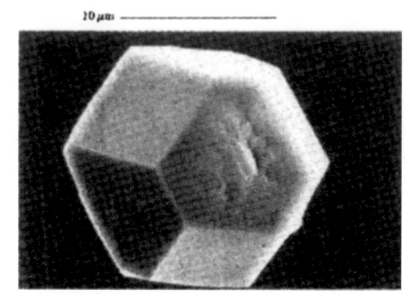

FIGURE 2.5   High-quality single-crystal diamond formed in the 2450-MHz discharge tube reactor. (After Chang, C.-P. et al., *J. Appl. Phys.* 63, 1744, 1988.)

2.2 SAPPHIRE

The crystal structure of sapphire (corundum) was first determined by Pauling and Hendricks (1925). The chemical formula is Al_2O_3. The crystal class is hexagonal system with rhombohedral class 3 m. The lattice constants are $a = 4.785$ Å and $c = 12.991$ Å. Figure 2.6 shows a schematic of the packing of O^{2-} ions (light circles) and Al^{3+} ions (black circles). The lattice constant a is equal to the distance of the O^{2-} ion from the c-axis passing through the Al^{3+} ions.

A vertical-pulling technique is used for the growth of highly perfect sapphire single crystals (Cockayne et al. 1967). Figure 2.7

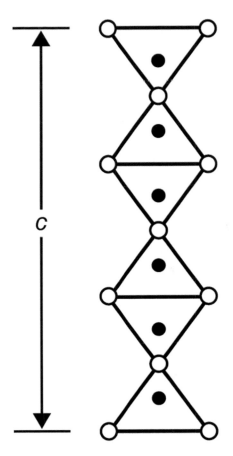

FIGURE 2.6 Schematic of the packing of O^{2-} (light circles) and Al^{3+} (black circles) ions in sapphire in the direction of the c-axis.

shows the vertical-pulling apparatus used for sapphire single-crystal growth. Pull rates of 6–50 mm/h and rotation rates of 0–200 rev/min were used.

A seeded vertical-gradient-freeze (VGF) method that yields Ti-doped laser-quality laser crystals with relatively low infrared absorption has been developed at Lincoln Laboratory (Sanchez et al. 1988). Figure 2.8 shows the apparatus for vertical-gradient-freeze growth of $Ti:Al_2O_3$ crystals. With a He atmosphere, the

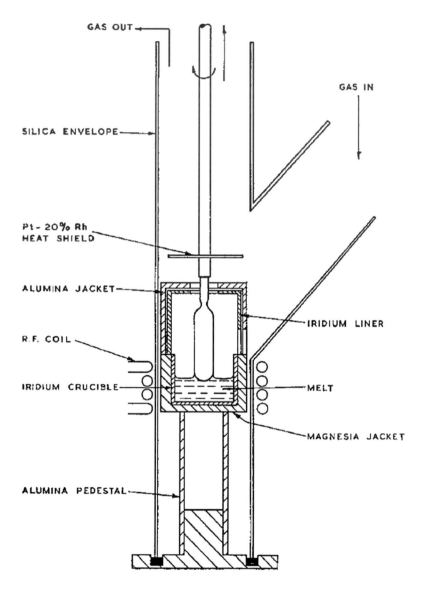

FIGURE 2.7   Vertical-pulling apparatus used for sapphire single-crystal growth. (After Cockayne, B. et al., *J. Materials Sci.* 2, 7, 1967.)

temperature gradient is about 15 °C/cm and the growth rate is about 2 mm/hr. Since the Ti^{3+} concentration in $Ti:Al_2O_3$ crystals grown by the VGF method varies with distance from the seed, laser rods are cut perpendicular to the growth axis in order to minimize the variation in concentration along their lengths.

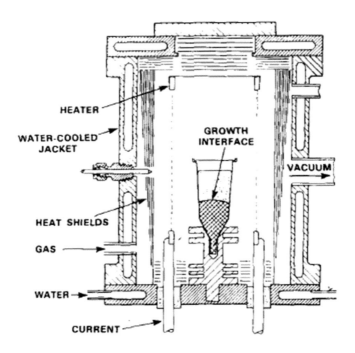

FIGURE 2.8 Apparatus for vertical-gradient-freeze growth of $Ti:Al_2O_3$ crystals. (After Sanchez, A. et al., *IEEE J. Quantum Electron.* 24, 995, 1988.)

CHAPTER **3**

Mechanical Properties of Diamond

3.1 HARDNESS

Diamond is the hardest known material, with a hardness of 10 Mohs, as it is composed of 98.9% ^{12}C carbon. Ramdas and coworkers have measured diamonds made of ^{13}C (Ramdas et al. 1993). Elastic constants for ^{13}C diamond are ~0.5% higher than the corresponding values for natural diamond. Diamond composed of ^{13}C is the *hardest known material*. Diamond is four times harder than sapphire, which has a hardness of 9 Mohs. The Vickers hardness of CVD diamond is 1×10^4 kg/mm^2 (Diamond Materials 2018). Hardness measurements have been reported to be 105 GPa using a nanohardness tester and conventional Vickers microhardness tester (Chowdhury et al. 2004).

3.2 TENSILE STRENGTH

Tensile strength is the greatest longitudinal stress a substance can bear without tearing apart. The tensile strength of diamond is the highest in the [100] crystal direction, smaller in the [110] direction, and smallest in the [111] direction, which is along the cube diagonal. The maximum tensile strengths have been calculated to be 225, 130, and 90 GPa in the [100], [110], and [111] directions, respectively (Telling et al. 2000). The precise tensile strength of diamond is unknown. Tensile strengths up to 60 GPa have been observed.

3.3 COMPRESSIVE STRENGTH

Compressive strengths of diamond have been calculated to be 223.1, 469.0, and 470.4 GPa in the [100], [110], and [111] directions, respectively (Luo et al. 2010). The compressive strength of diamond has been measured to be 1.6 Mbar (160 GPa) using a diamond tip of 2 μm radius on a flat diamond anvil (Ruoff and Wanagel 1977).

3.4 YOUNG'S MODULUS

Young's modulus is named after the nineteenth-century British scientist Thomas Young. Young's modulus E defines the relationship between stress (force/area) and strain (fractional change in length), which is given by

$$E = \frac{F/A}{\Delta L/L_0} \tag{3.1}$$

where F is the force, A is the cross-sectional area through which the force is applied, ΔL is the change in the length, and L_0 is the original length. E depends on the direction of the crystallographic plane identified by Miller indices (hkl) (Turley and Sines 1971).

$$E_{(hkl)} = \left(s_{12} + \frac{1}{2} s_{44} + \frac{S}{\Omega} \right)^{-1} \tag{3.2}$$

where

$$S = (s_{11} - s_{12}) - \frac{1}{2} s_{44} \tag{3.3}$$

where s_{11}, s_{12}, and s_{44} are the elastic compliance constants, and

$$\Omega = \frac{3}{4}(1-a) - b\sin(2\theta) - c\cos(2\theta) - d\sin(4\theta) - e\cos(4\theta) \quad (3.4)$$

where θ specifies a direction in the plane and a, b, c, d, and e are coefficients that depend on the two angles. Grimsditch and Ramdas have measured elastic compliances s_{11}, s_{12}, and s_{44} of single-crystal diamond using Brillouin scattering equal to 0.9524, −0.0991, and 1.7331 TPa^{-1}, respectively (Grimsditch and Ramdas 1975). Using these values of s_{11}, s_{12}, and s_{44}, the value of S is determined to be 0.1849 TPa^{-1}. The coefficients a, b, c, d, and e are listed in Table 3.1 for the highest-symmetry cubic crystallographic planes (100), (110), and (111) (Turley and Sines 1971).

TABLE 3.1   Values of the Coefficients a, b, c, d, and e for the (100), (110), and (111) Planes

(hkl)	a	b	c	d	e
(100)	0	0	0	0	−1/4
(110)	1/4	0	−1/4	0	−3/16
(111)	1/3	0	0	0	0

Ω for the (111) plane is equal to 1/2, giving

$$E_{111} = \left(s_{12} + \frac{1}{2}s_{44} + 2 \right)^{-1} \quad (3.5)$$

Using the values −0.0991, 1.7331, and 0.1849 TPa^{-1} for s_{12}, s_{44}, and S, one obtains a value of 871 GPa for $E_{(111)}$. Values of $E_{(100)}$ and $E_{(110)}$ are more complex (Klein and Cardinale 1993).

Young's modulus exhibits relatively little anisotropy. The generally accepted value of 1050 GPa for E represents a minimum value.

3.5 ELASTIC CONSTANTS

Elastic constants of diamond have been determined by several researchers, including Bhagavantam and Bhimasenachar using the measurement of ultrasonic velocities (1946), Grimsditch and Ramdas using Brillouin scattering (1975), and Wang and Ye using ab initio calculations (2003). Table 3.2 lists the values of the elastic constants c_{11}, c_{12}, and c_{44} of diamond determined by the above authors.

TABLE 3.2   Values of the Elastic Constants c_{11}, c_{12}, and c_{44} of Diamond

Elastic Constant	Bhagavantam and Bhimasenachar	Grimsditch and Ramdas	Wang and Ye
c_{11} (dyn/cm^2)	9.5×10^{12}	$10.764 \pm 0.002 \times 10^{12}$	$10.996 \pm 0.21 \times 10^{12}$
c_{12} (dyn/cm^2)	3.9×10^{12}	1.252×10^{12}	$1.428 \pm 0.22 \times 10^{12}$
c_{44} (dyn/cm^2)	4.3×10^{12}	5.774 ± 0.014	$5.870 \pm 0.094 \times 10^{12}$

CHAPTER 4

Mechanical Properties of Sapphire

4.1 HARDNESS

Hardness is measured in terms of Mohs, Vickers, and Knoop. The hardness of sapphire was determined to be 9 Mohs compared with that of 10 Mohs for diamond, which is the hardest material of all. The Vickers hardness number for synthetic sapphire is 2720 (Taylor 1949). The Vickers hardness of synthetic sapphire is 17.4 GPa parallel to the c-axis and 15.6 GPa perpendicular to the c-axis (Haney and Subash 2011). The load-independent Knoop microhardness is 1170 kg/mm^2 (Kaji et al. 2002). Figure 4.1 shows the Knoop microhardness of synthetic sapphire on the basal plane (0001) and its variation with crystallographic orientation and indentation test load (Kaji et al. 2002).

4.2 TENSILE STRENGTH

Tensile strength is the greatest longitudinal stress a substance can bear without tearing apart. Figure 4.2 shows the tensile strength of sapphire measured along the a- and c-axes at temperatures between 20°C and 800°C (Schmid and Harris 1998).

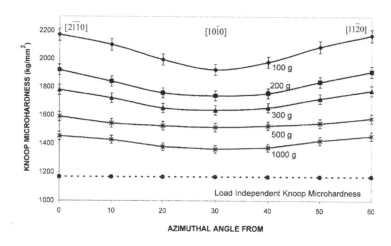

FIGURE 4.1   Knoop microhardness of synthetic sapphire on the basal plane (0001) and its variation with crystallographic orientation and indentation test load. (After Kaji, M. et al., *J. Am. Ceram. Soc.* 85, 415, 2002.)

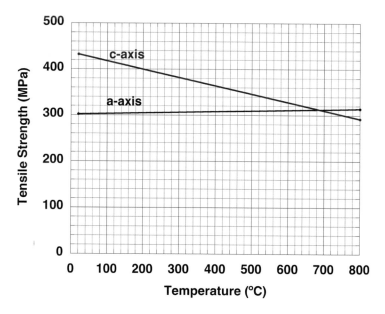

FIGURE 4.2   Tensile strength of sapphire along the *a*- and *c*-axes at temperatures between 20°C and 800°C.

4.3 YOUNG'S MODULUS

Figure 4.3 shows the variation of the Young's modulus of ruby and sapphire as a function of temperature.

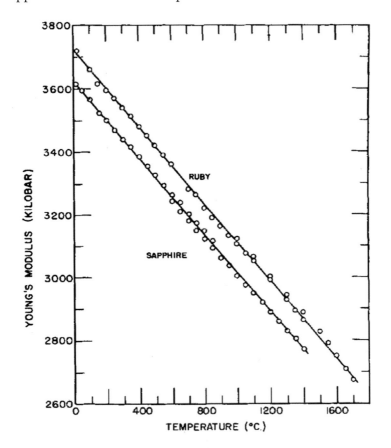

FIGURE 4.3 Young's modulus of ruby and sapphire as a function of temperature. (After Wachtman, J. B., Jr., and D. G. Lam, Jr., *J. Am. Ceramic Soc.* 42, 254, 1959.)

The room-temperature values of the Young's modulus for ruby and sapphire are 3730 kilobar (373 GPa) and 3620 kilobar (362 GPa), respectively, which are lower by a factor of ~3 compared to that of 1050 GPa for diamond.

4.4 ELASTIC CONSTANTS

The elastic constants of sapphire have been determined by several researchers, including Wachtman et al. using a resonance technique (1960), Bernstein using pulse-echo and cw resonance measurements (1963), and Tefft using a resonance technique over the temperature range 80–900 K (1966). Table 4.1 lists elastic constants c_{11}, c_{33}, c_{44}, c_{12}, c_{13}, and c_{14} of sapphire at room temperature as determined by Wachtman et al. and Bernstein.

TABLE 4.1   Values of the Elastic Constants c_{11}, c_{33}, c_{44}, c_{12}, c_{13}, and c_{14} of Sapphire at Room Temperature

Elastic Constant (10^{12} dyn/cm²)	Wachtman et al.	Bernstein
c_{11}	4.968 ± 0.018	4.911
c_{33}	4.981 ± 0.014	4.911
c_{44}	1.474 ± 0.002	1.461
c_{12}	1.636 ± 0.018	1.669
c_{13}	1.109 ± 0.022	1.147
c_{14}	-0.235 ± 0.003	−0.233

CHAPTER 5

Thermal Properties of Diamond

5.1 THERMAL EXPANSION

Several x-ray measurements of thermal expansion of diamond have been reported, including those of Krishnan (1946), Thewlis and Davey (1956), Novikova (1961), and Sokhor and Vitol (1970). Figures 5.1 and 5.2 show plots of the coefficient of linear thermal expansion α vs T obtained using the data of Thewlis and Davey for gem-quality and industrial diamonds, respectively (1956). α for industrial diamond shows a minimum of $\sim$255 K.

According to Gruneisen's law (1926), α is given by

$$\alpha = \frac{\gamma \chi_0 C_V}{3V_0} \tag{5.1}$$

where γ is the Gruneisen number, χ_0 is the compressibility, and V_0 is the atomic volume. γ was supposed to be a constant. However, it was found to vary with temperature, as shown in Figure 5.3, obtained using the data of Thewlis and Davey for α of gem-quality diamond, and the Debye model for C_V yields θ_D equal to 1880 K.

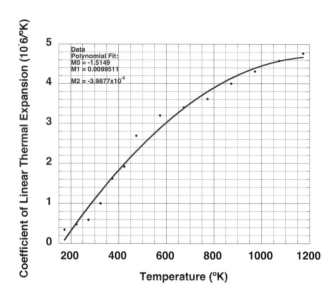

FIGURE 5.1 Plot of the coefficient of linear thermal expansion vs temperature for gem-quality diamond.

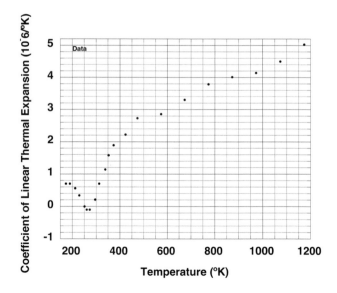

FIGURE 5.2 Plot of the coefficient of linear thermal expansion vs temperature for industrial diamond.

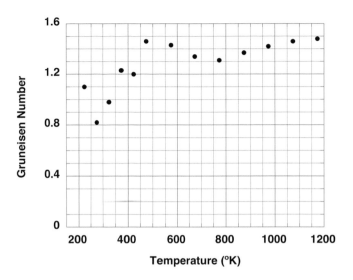

FIGURE 5.3   Plot of the Gruneisen number vs temperature for diamond.

5.2 SPECIFIC HEAT

The specific heat of diamond is due to phonons. In the Debye model, the molar specific heat at constant volume C_v is given by

$$C_v = 9N_A k_B \left(\frac{T}{\theta_D}\right)^3 \int_0^{\theta_D/T} \frac{x^4 e^x}{(e^x - 1)^2} \, dx \qquad (5.2)$$

where N_A is Avogadro's number, k_B is the Boltzmann constant, T is the temperature in K, and θ_D is the Debye temperature. A value of 1880 ± 10 K has been determined for θ_D by Victor based on measurements of specific heat in the temperature range 300–1100 K (1962). Raman has also calculated values of C_V based on the spectroscopic behavior of diamond (1957). Figure 5.4 provides a comparison of the calculated values of C_V obtained from Equation 5.2 using the value of 1880 K for θ_D with those calculated by Raman in the 300–1100 K temperature range. The two sets of values of C_V for the Debye model and Raman calculations are in good agreement.

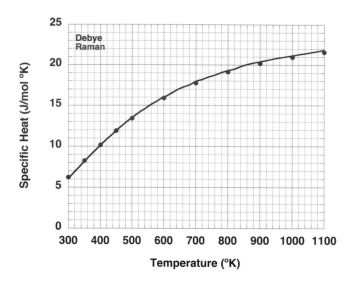

FIGURE 5.4   Comparison of the values of C_V calculated from Equation 5.2 with those calculated by Raman.

The Debye temperature is found to vary in the temperature range 12.8–277 K, as shown in Figure 5.5 (Desnoyehs and Morrison 1958). Therefore, we will use the C_V values calculated by Raman for comparing the data for temperatures both below and above room temperature.

Figure 5.6 shows the below-room-temperature specific heat C_P measurements of Desnoyehs and Morrison (1958) and above-room-temperature measurements of Victor (1962) along with the C_V values calculated by Raman. There is reasonably good agreement between the measured C_P values and the calculated C_V values. The difference between C_P and C_V is expected to be small.

For low temperatures $T < {\sim}\theta_D/20$, C_V for the Debye model scales as T^3. This implies that $C_V^{1/3}$ should scale as T. Figure 5.7 shows the measured variation $C_V^{1/3}$ in the 12.8–100 K temperature range (Desnoyehs and Morrison 1958) compared with the values of $C_V^{1/3}$ calculated by Raman (1957).

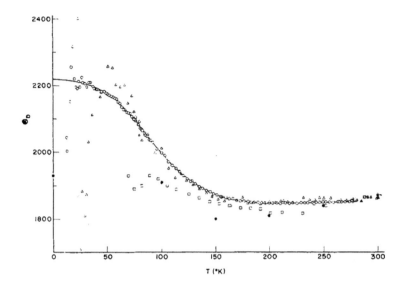

FIGURE 5.5 Variation of θ_D of diamond with temperature. (After Desnoyehs, J. E., and J. A. Morrison, *Philos. Mag.* 36, 42, 1958.)

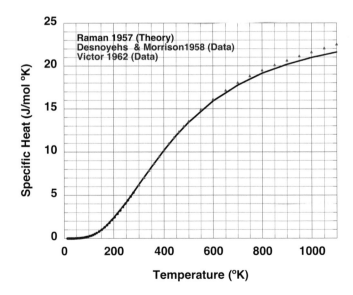

FIGURE 5.6 Specific heat C_P measurements of diamond compared with the C_V values of Raman.

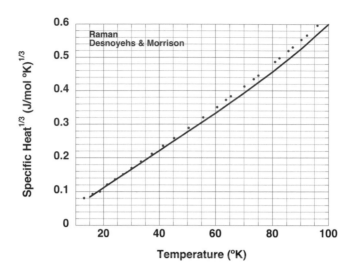

FIGURE 5.7 Plot of the cube root of specific heat of diamond vs temperature.

5.3 THERMAL CONDUCTIVITY

The thermal conductivity κ of diamond is due to phonons and is given by

$$\kappa = \frac{1}{3} C_v v l \qquad (5.3)$$

where v is the speed of the phonons and l is the mean free path of the phonons. Figure 5.8 shows the temperature dependence of κ for type IIa, IIb, and I diamonds measured by Berman et al. (1956). At low temperatures, $\kappa \sim T^3$, which is due to the temperature dependence of C_v. At high temperatures, $\kappa \sim 1/T$, which is due to the temperature dependence of l. The κ curves for type IIb and I diamonds are lower than those for type IIa diamonds. This is also due to the lower values of l due to the scattering of phonons by substitutional boron atoms in type IIb and nitrogen atoms in type I diamonds.

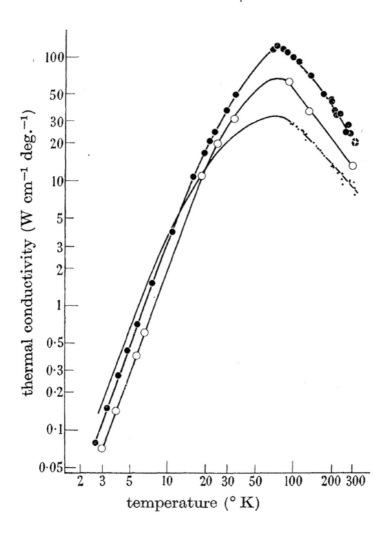

FIGURE 5.8   Thermal conductivity of type IIa (●), type IIb (○), and type I (●) diamonds. (After Berman, R. et al., *Proc. Roy. Soc. (London)* A237, 344–354, 1956.)

Figure 5.9 shows the temperature dependence of the thermal conductivity of CVD diamonds (CVD Diamond Booklet 2018) compared with that of copper (red curve).

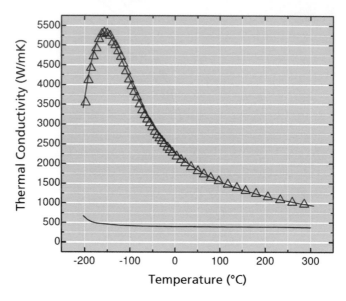

FIGURE 5.9   Thermal conductivity of CVD diamond compared with that of copper (red curve). (After Diamond Materials, "The CVD diamond booklet," 2018.)

CHAPTER 6

Thermal Properties of Sapphire

6.1 THERMAL EXPANSION

Figure 6.1 shows the variation of the lattice constants a and c of sapphire with temperature obtained from precision x-ray lattice parameter measurements (Yim and Paff 1974).

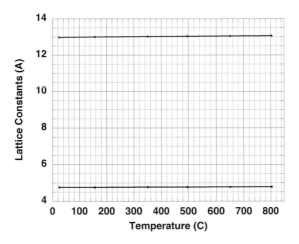

FIGURE 6.1   Plot of the lattice constants a (black curve) and c (red curve) of sapphire vs temperature.

Figure 6.2 shows the variation of the coefficients of thermal expansion of sapphire with temperature (Yim and Paff 1974).

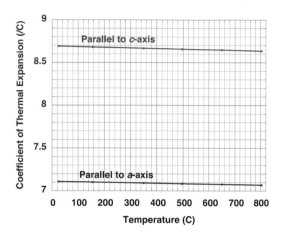

FIGURE 6.2   Plot of the coefficient of thermal expansion of sapphire vs temperature.

6.2 SPECIFIC HEAT

Figure 6.3 shows the specific heat of sapphire as a function of temperature from 0 to 1200 K (Furukawa et al. 1956).

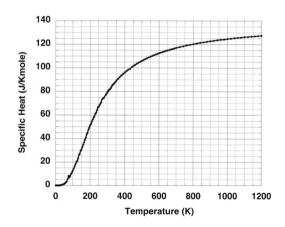

FIGURE 6.3   Specific heat of sapphire as a function of temperature from 0 to 1200 K.

The specific heat of sapphire has been measured from 2 to 25 K by Fugate and Swenson (1969). Figure 6.4 compares the data of Fugate and Swenson with that of Furukawa et al. (1956).

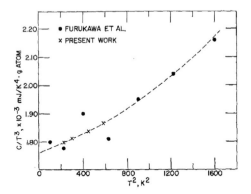

FIGURE 6.4 Specific heat of sapphire from 2 to 25 K measured by Fugate and Swenson compared with that of Furukawa et al. (After Fugate, R. Q., and C. A. Swenson, *J. Appl. Phys.* 40, 3034, 1969.)

6.3 THERMAL CONDUCTIVITY

Figure 6.5 shows measurements of the thermal conductivity of sapphire over the temperature range ~120–750 K (Cahill et al. 1998).

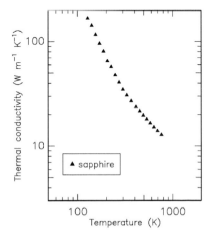

FIGURE 6.5 Thermal conductivity of sapphire. (After Cahill, D. G. et al., *J. Appl. Phys.* 83, 5783, 1998.)

CHAPTER 7

Optical Properties of Diamond

7.1 TRANSMISSION RANGE

Figure 7.1 shows the transmission spectrum of CVD diamond for the spectral range 0–100 μm (CVD Diamond Booklet 2018). This spectrum has not been corrected for reflection losses.

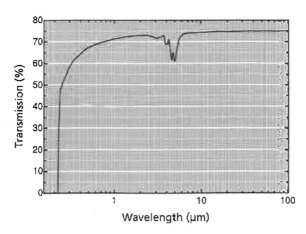

FIGURE 7.1   Transmission spectrum of CVD diamond. (After Diamond Materials, "The CVD Diamond Booklet," 2018.)

7.2 REFRACTIVE INDEX

Figure 7.2 shows the refractive index of diamond using the Sellmeier equation (Peter 1923)

$$n = \left(1 + \frac{0.3306\lambda^2}{\lambda^2 - 0.175^2} + \frac{4.3356\lambda^2}{\lambda^2 - 0.106^2}\right)^{1/2} \tag{7.1}$$

where λ is the wavelength in μm.

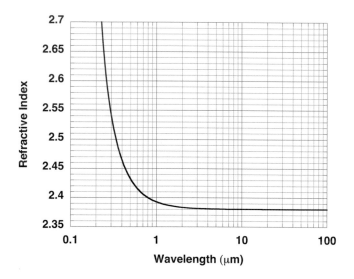

FIGURE 7.2   Refractive index of diamond.

The change in the refractive index of diamond with temperature at 632.8 nm is given by (Patterson et al. 1995)

$$\frac{dn}{dT} = c_0 + c_1 T + c_2 T^2 + c_3 T^3 \tag{7.2}$$

where T is the temperature in °C, $c_0 = 7.446 \times 10^{-6}/°C$, $c_1 = 1.071 \times 10^{-7}/°C^2$, $c_2 = -8.832 \times 10^{-11}/°C^3$, and $c_3 = 2.911 \times 10^{-14}/°C^4$. Figure 7.3 shows a plot of dn/dT vs T using Equation 7.2.

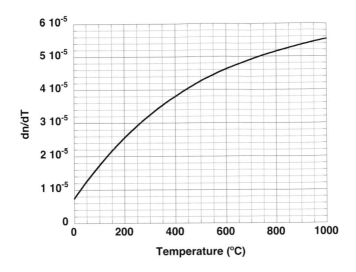

FIGURE 7.3   Plot of *dn/dT* vs *T* for diamond.

The high refractive index (>2.41) of diamond in the visible region is responsible for its brilliance/luster in an "ideal-cut" diamond. There are three ideal cuts in common use: the American Standard developed by Marcel Tolkowsky in 1919, Practical Fine Cut used in Germany and other European countries, and Scandinavian Standard. Diamond proportions for the American Standard, Practical Fine Cut, and Scandinavian Standard are listed in Table 7.1.

TABLE 7.1   Diamond Proportions for the American Standard, Practical Fine Cut, and Scandinavian Standard

Ideal Cut	Crown Height	Pavilion Depth	Table Diameter	Crown Angle	Pavilion Angle
American Standard	16.2%	43.1%	53.0%	34.5°	40.75°
Practical Fine Cut	14.4%	43.2%	56.0%	33.2°	40.8°
Scandinavian Standard	14.6%	43.1%	57.5%	34.5°	40.75°

Source: Hemphill, T. S. et al. *Gems Gemol.* 34, 158, 1998.

Light incident on the ideal-cut diamond undergoes two total internal reflections before it emerges out of the diamond. The brilliance grades are 99.5%, 99.95%, and 99.5% for the American Standard, Practical Fine Cut, and Scandinavian Standard.

7.3 ABSORPTION SPECTRUM

Figure 7.4 shows the absorption spectrum of a natural white diamond (red trace) and a natural yellow diamond (blue trace) (PerkinElmer).

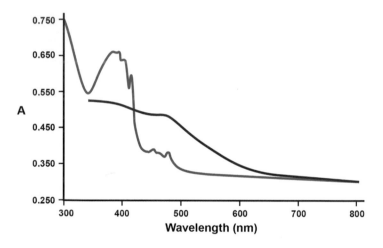

FIGURE 7.4 Absorption spectrum of a natural white diamond (red trace) and a natural yellow diamond (blue trace). (After PerkinElmer, "Acquisition of high-quality transmission spectra of ultra-small samples using the Lambda 950 UV/Vis/NIR and Lambda 850 UV/Vis spectrophotometers," Shelton, CT 06484-4794 USA.)

Figure 7.5 shows the infrared (IR) absorption spectrum of a type Ia diamond (Aggarwal et al. 2012).

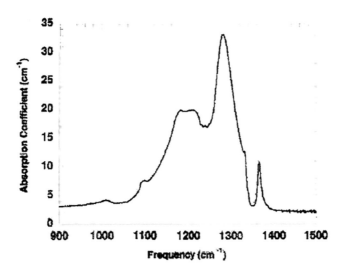

FIGURE 7.5 IR absorption spectrum of type 1a diamond. (After Aggarwal, R. L. et al., *Solid State Commun.* 152, 204, 2012.)

Absorption peaks are observed at 1010, 1090, ~1200, 1282, 1332, and 1365 cm^{-1}. The absorption peaks at 1090, ~1200, and 1282 cm^{-1} correspond to nitrogen in the A form (Clark et al. 1979; Woods et al. 1990). The absorption peak at 1365 cm^{-1} called B' is due to plate defects on the (001) plane (Evans and Phaal 1962; Woods et al. 1990).

CHAPTER 8

Optical Properties of Sapphire

8.1 TRANSMISSION AND ABSORPTION

Figure 8.1 shows the transmission spectrum of a 2-mm-thick synthetic sapphire window (Kopchatov 2018).

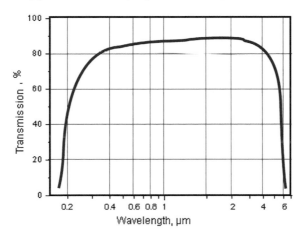

FIGURE 8.1   Transmission spectrum of 2-mm-thick synthetic sapphire window. (After Kopchatov, V., "Synthetic sapphire," J. S. Tydex Co., St. Petersburg, Russia, 2018.)

Because of the wide optical transmission range, synthetic sapphire is used for UV, VIS, and NIR optics. Figure 8.2 shows the absorption spectrum of blue sapphire (PerkinElmer).

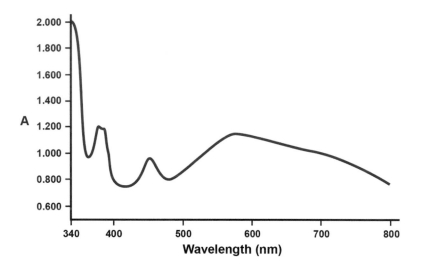

FIGURE 8.2   Absorption spectrum of blue sapphire. (After PerkinElmer, "Acquisition of high-quality transmission spectra of ultra-small samples using the Lambda 950 UV/Vis/NIR and Lambda 850 UV/Vis spectrophotometers," Shelton, CT 06484-4794 USA.)

8.2 REFRACTIVE INDEX

Figure 8.3 shows the ordinary and extraordinary refractive indices n_O and n_E of synthetic sapphire (Jeppesen 1958).

The temperature coefficients for n_O and n_E at 589.3 nm have been measured to be 13.6×10^{-6}/K and 14.7×10^{-6}/K, respectively. Figure 8.4 shows the birefringence $(n_E - n_O)$ of synthetic sapphire (Jeppesen 1958). The temperature coefficient of birefringence at 590.0 nm has been measured to be 1.1×10^{-6}/K (Jeppesen 1958).

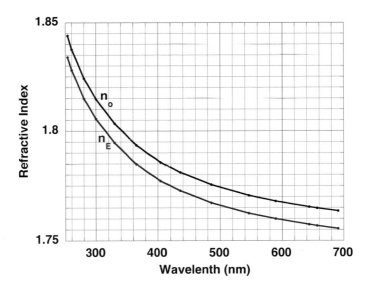

FIGURE 8.3   Refractive indices n_O and n_E of synthetic sapphire.

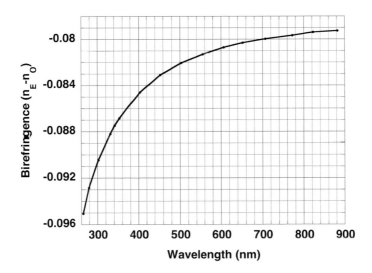

FIGURE 8.4   Birefringence ($n_E - n_O$) of synthetic sapphire.

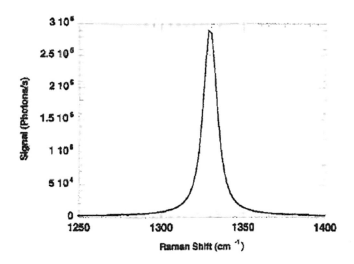

FIGURE 9.3 First-order Raman spectrum of a 0.9-mm-thick (111) diamond. (After Aggarwal, R. L. et al., *Solid State Commun.* 152, 204, 2012.)

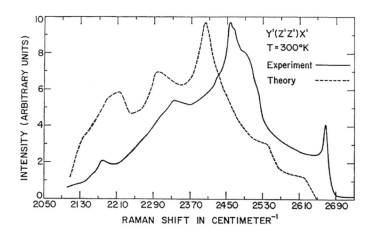

FIGURE 9.4 Comparison of the calculated and experimental results for the room-temperature second-order Raman scattering in diamond; the strongest peak in the calculation has been normalized to the strongest peak in the experimental spectrum. (After Solin, S. A., and A. K. Ramdas, *Phys. Rev. B* 1, 1687, 1970.)

experimental results for the room-temperature second-order Raman scattering in diamond (Solin and Ramdas 1970).

9.2 BRILLOUIN SCATTERING

Brillouin scattering in crystals is inelastic scattering of light caused by acoustic phonons. It was first predicted by Leon Brillouin (1922). However, Leonid Mandelstam is believed to have recognized the possibility of such scattering as early as 1918, but he published his idea only in 1926.

Brillouin scattering of diamond was first reported by Krishnan (1947a) and later on by Chandrasekharan (1950), Krishnan et al. (1958), and Grimsditch and Ramdas (1975). The Brillouin frequency shift is given by (Grimsditch and Ramdas 1975)

$$\nu_B = 2n\left(\frac{v_s}{c}\right)\nu_L \sin\left(\frac{\theta}{2}\right) \qquad (9.3)$$

where n is the refractive index, v_s is the velocity of the sound wave, c is the speed of light, ν_L is the frequency of the excitation laser, and θ is the scattering angle. The sound velocity v_s is given by

$$v_s = \sqrt{\frac{X}{\rho}} \qquad (9.4)$$

where X is an appropriate combination of elastic moduli c_{11}, c_{12}, and c_{44}, which have been determined to be equal to $10.764 \pm 0.002 \times 10^{12}$, $1.252 \pm 0.023 \times 10^{12}$, and $5.774 \pm 0.014 \times 10^{12}$ dyn/cm², respectively (Grimsditch and Ramdas 1975). ρ is the density equal to 3.515 g/cm³. X is equal to c_{11} and c_{44}, respectively, for the longitudinal and transverse phonons traveling along the [100] direction. In this case, the values of v_s are 1.75×10^6 and 1.28×10^6 cm/s, respectively, for the longitudinal and transverse phonons. The values of the Brillouin frequency shifts ν_B are 1.23×10^{11} Hz (4.1 cm⁻¹) and 9.00×10^{10} Hz (3.0 cm⁻¹), respectively, for the longitudinal and transverse phonons using the values of 2.45 for n, 6.15×10^{14} Hz for ν_L, and 90° for θ.

Figure 9.5 shows the Brillouin spectrum of diamond at room temperature for 488.0-nm laser excitation propagating along the x' direction and scattered along the y' direction. In the labels VH and VV, the first and second letters denote the polarization (V: vertical, H: horizontal) of the incident and scattered light with respect to the horizontal (001) scattering plane. T (transverse) and L (longitudinal) give the polarization characteristics of the phonons propagating parallel to the [100] direction (Grimsditch and Ramdas 1975).

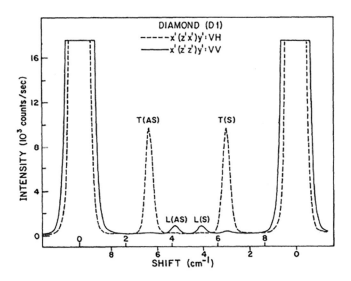

FIGURE 9.5   Brillouin spectrum of diamond at room temperature for the VH and VV polarizations. (After Grimsditch, M., and A. K. Ramdas, *Phys. Rev. B* 11, 3139, 1975.)

The intensities of the Brillouin components for the L and T phonons are given by $4(p_{12})^2/c_{11}$ and $2(p_{44})^2/c_{44}$, respectively, where $p_{12} = 0.043$ and $p_{44} = -0.172$ are the elasto-optic constants. Using these values of p_{12} and p_{44}, the intensities of the L and T phonons are given by 0.00068 and 0.0125. Hence, the intensity of the L phonons is much weaker than that of the T phonons, as shown in Figure 9.5.

CHAPTER 10

Light Scattering of Sapphire

10.1 RAMAN SCATTERING

Raman scattering of sapphire was first reported by Krishnan in 1947 (1947b). Later measurements of Raman scattering of sapphire include Porto and Krishnan (1967) and Watson et al. (1981). Pressure dependence of the Raman-active modes in sapphire also has been reported (Watson et al. 1981). Temperature dependence of the Raman scattering of sapphire has been reported recently (Thapa et al. 2017).

There are two Raman modes of A_1g symmetry and five modes of E_g symmetry. The two A_1g modes are observed at 417.4 and 644.6 cm^{-1}, and the five E_g modes are observed at 378.7, 430.2, 448.7, 576.7, and 750.0 cm^{-1} (Watson et al. 1981). The Raman tensors of sapphire are (Loudon 1964):

$$A_{1g} = \begin{vmatrix} a & 0 & 0 \\ 0 & a & 0 \\ 0 & 0 & b \end{vmatrix} \qquad (10.1)$$

and

$$E_g = \begin{vmatrix} c & -c & -d \\ -c & c & d \\ -d & d & 0 \end{vmatrix} \tag{10.2}$$

There are only independent Raman tensor components: $\alpha_{xx} = a$, $\alpha_{zz} = b$, $\alpha_{xy} = -c$, and $\alpha_{xz} = -d$. Figure 10.1 shows the Raman spectra of sapphire (Watson et al. 1981).

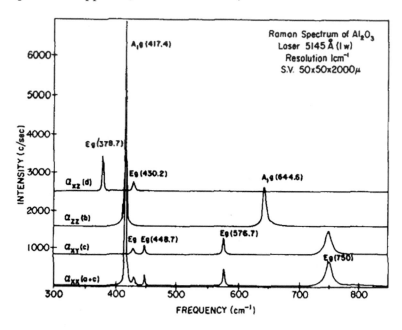

FIGURE 10.1 Raman spectra of sapphire. (After Watson, G. H. et al., *J. Appl. Phys.* 52, 956, 1981.)

The Raman cross-section of the 417.4-cm^{-1} mode has been determined to be 1.59×10^{-30} cm^2/sr unit cell or 2.00×10^{-29} cm^2/unit cell (Watson et al. 1981). Figure 10.2 shows the pressure dependence of the Raman-active modes of sapphire (Watson et al. 1981).

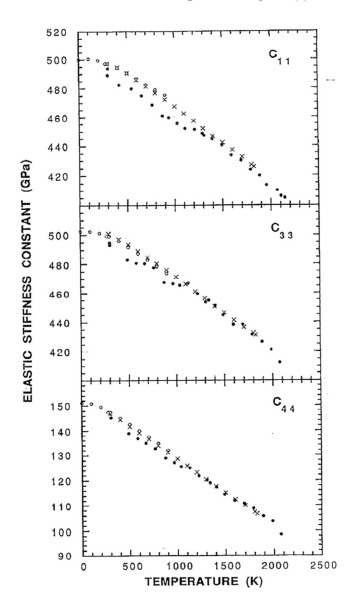

FIGURE 10.5 Temperature dependence of the diagonal elastic constants of single-crystal sapphire. (After Zouboulis, E. S., and M. Grimsditch, *J. Appl. Phys.* 70, 772, 1991.)

CHAPTER 11

Sapphire Lasers

11.1 CHROMIUM-DOPED SAPPHIRE LASER

The chromium-doped sapphire (ruby) laser was the first laser discovered by T. H. Maiman in 1960 at the Hughes Research Laboratory (Maiman 1960). At the time of its discovery, the first laser was dubbed a "death ray" by some and a "solution looking for a problem" by others. Today the laser touches our lives in countless ways—from supermarket scanners to DVD players, from cosmetic surgery to state-of-the-art medical advances, from connecting people through the internet to keeping our communities secure (SPIE 2017).

Ruby is chromium-doped Al_2O_3 ($Cr:Al_2O_3$). Chromium gives the ruby its red color. Components of the first ruby laser are (Laserfest 2010): (i) power supply, (ii) switch, (iii) 100% reflective mirror, (iv) quartz flash tube, (v) ruby crystal, (vi) 95% reflective mirror, and (vii) polished aluminum reflective cylinder.

Figure 11.1 shows the energy-level diagram of the ruby laser at 694.3 nm due to the R_1 transition from ^2E to $^4A_2$.

11.2 TITANIUM-DOPED SAPPHIRE LASER

Laser operation from the transition-metal ion Ti^{3+} in sapphire was first observed by Peter Moulton at the MIT Lincoln Laboratory

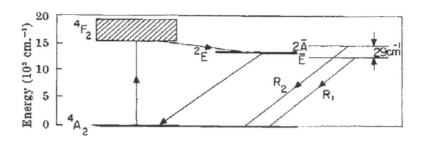

FIGURE 11.1 Energy-level diagram of ruby laser at 693.4 nm. (After Maiman, T. H., *Nature* 187, 493, 1960.)

in 1982 (Moulton 1982). This represents the first use of Ti^{3+} as a laser-active ion and only the second example of sapphire as a host crystal, the first being the Cr^{3+} ion in the ruby laser. In preliminary experiments, pulsed tunable laser operation from 718 to 770 nm was obtained by optical pumping into the $Ti:Al_2O_3$ absorption band, and peak output powers of ~5 kW in a 500-ns-long pulse were generated. Lemoff and Barty reported the generation of 804-nm pulses with a duration as short as 20 fs and with peak powers as high as 500 kW from a regeneratively initiated, self-mode-locked Ti:sapphire laser (1992). Asaki et al. reported the generation of 780-nm 11-fs pulses with a bandwidth of 62 nm and average power of 500 mW (1993).

There is no excited state absorption in the $Ti:Al_2O_3$ laser. Consequently, the $Ti:Al_2O_3$ laser is expected to operate over the entire fluorescence region, 650–900 nm. The $Ti:Al_2O_3$ laser has a much higher gain cross-section, which should allow generation of higher peak powers and shorter pulses in Q-switched mode. Figure 11.2 shows fluorescence lifetime versus temperature for the ^2E to $^2T_2$ transition in $Ti:Al_2O_3$ for a sample with 0.1 wt. % Ti.

The rapid reduction in the fluorescence lifetime at high temperatures is characteristic of fluorescence quenching due to multiphonon nonradiative decay.

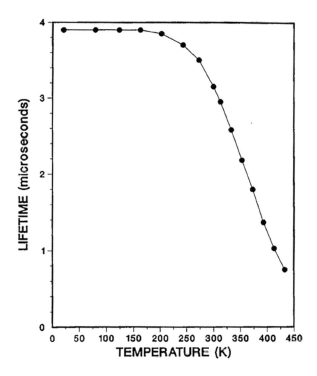

FIGURE 11.2 Fluorescence lifetime vs temperature for the ^2E to $^2T_2$ transition in Ti:Al$_2$O$_3$. (After Moulton, P. F., *J. Opt. Soc. Am. B* 3, 125, 1986.)

Room-temperature continuous-wave (cw) operation of a Ti:Al$_2$O$_3$ laser was first reported by Sanchez et al. in 1986 at 770 nm by pumping with an Ar-ion laser, using all the lines of the blue-green region (Sanchez et al. 1986). Figure 11.3 shows a schematic of the experimental setup of the room-temperature cw Ti:Al$_2$O$_3$ laser showing the Ar-ion pump laser and the three-mirror folded cavity.

Figure 11.4 shows the Ti:Al$_2$O$_3$ laser output at 770 nm vs incident pump for two output couplers with transmittance T equal to 0.7% and 4.9%.

Using the measured values of the slope quantum efficiency, the value of the internal quantum efficiency was determined to be $64 \pm 10\%$.

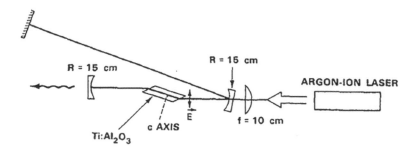

FIGURE 11.3   Schematic of the experimental setup of the room-temperature cw Ti:Al$_2$O$_3$ laser at 770 nm showing the Ar-ion pump laser and three-mirror folded cavity. (After Sanchez, A. et al., *Opt. Lett.* 11, 363, 1986.)

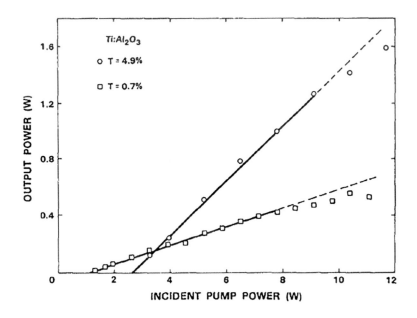

FIGURE 11.4   Ti:Al$_2$O$_3$ laser output at 770 nm versus incident pump power for two couplers with transmittance *T* equal to 0.7% and 4.9%. (After Sanchez, A. et al., *Opt. Lett.* 11, 363, 1986.)

References

Aggarwal, R. L., L. W. Farrar, S. K. Saikin, X. Andrade, A. Aspuru-Guzik, and D. L. Polla, "Measurement of the absolute Raman cross section of the optical phonon in type 1a natural diamond," *Solid State Commun.* 152, 204, 2012.

Asaki, M. T., C.-P. Huang, D. Garvey, J. Zhou, H. C. Kapteyn, and M. M. Murnane, "Generation of 11-fs pulses from a self-mode-locked Ti:sapphire laser," *Opt. Lett.* 18, 977, 1993.

Berman, R., E. L. Foster, and J. M. Ziman, "The thermal conductivity of dielectric crystals: The effect of isotopes," *Proc. Roy. Soc. (London)* A237, 344–354, 1956.

Bernstein, B. T., "Elastic constants of synthetic sapphire at 27°C," *J. Appl. Phys.* 34, 169, 1963.

Bhagavantam, S., "Relation of Raman effect to crystal structure and properties of diamond," *Indian J. Phys.* 5, 169, 1930.

Bhagavantam, S., and J. Bhimasenachar, "Elastic constants of diamond," *Proc. Royal Soc. (London)* A187, 381, 1946.

Brillouin, L., "Diffusion of light and x-rays by a transparent homogeneous body," *Ann. Phys. (Paris)* 17, 88, 1922.

Bundy, F. P., H. T. Hall, H. M. Strong, and R. H. Wentorf, "Man-made diamonds," *Nature* 176, 51 1955.

Cahill, D. G., S.-M. Lee, T. I. Selinder, "Thermal conductivity of κ-Al_2O_3 and α-Al_2O_3 wear-resistant coatings," *J. Appl. Phys.* 83, 5783, 1998.

Chandrasekharan, V., "Thermal scattering of light in crystals Part II. Diamond," *Proc. Ind. Acad. Sci. A* 32, 379, 1950.

Chang, C.-P., D. L. Flamm, D. E. Ibbotson, and J. A. Mucha, "Diamond crystal growth by plasma chemical vapor deposition," *J. Appl. Phys.* 63, 1744, 1988.

Chowdhury, S., E. D. Barra, and M. T. Laugier, "Study of mechanical properties of CVD diamond on SiC substrates," *Diam. Relat. Mater.* 13, 1625, 2004.

Moulton, P., "Ti-doped sapphire: Tunable solid-state laser," *Optics News*, November/December, 9, 1982.

Moulton, P. F., "Spectroscopic and laser characteristics of Ti:Al$_2$O$_3$," *J. Opt. Soc. Am. B* 3, 125, 1986.

Nayar, P. G. N., "Temperature variation of the Raman frequency of diamond," *Proc. Indian Acad. Sci.* 13, 284, 1941.

Novikova, S. I., "Thermal expansion of diamonds between 25 and 750°K," *Sov. Phys. Solid State* 2, 1464, 1961.

Palik, E. D. (Ed.), *Handbook of Optical Constants of Solids*, Academic Press, 1985, p. 565.

Patterson, M. J., J. L. Margrave, R. H. Hauge, Z. Ball, and R. Sauerbrey, "Measurement of the change in the refractive index of diamond with temperature and of the etching rate of diamond by the KrF (248 nm) excimer laser," *Electrochem. Soc. Proc.* 95–4, 503–508, 1995.

Pauling, L., and S. B. Hendricks, "The crystal structures of hematite and corundum," *J. Am. Chem. Soc.* 47, 781, 1925.

PerkinElmer, "Acquisition of high-quality transmission spectra of ultra-small samples using the Lambda 950 UV/Vis/NIR and Lambda 850 UV/Vis spectrophotometers," Shelton, CT.

Peter, F., "Uber brechungsindizes und absorptionkonstanten des diamenten zwischen 644 und 226 mμ," *Z. Phys.* 15, 358, 1923.

Porto, S. P. S., and R. S. Krishnan, "Raman effect of corundum," *J. Chem. Phys.* 47, 1009, 1967.

Raman, C. V., "A new radiation," *Indian J. Phys.* 2, 387, 1928.

Raman, C. V., "The diamond," *Proc. Indian Acad. Sci. Sect.* A44, 99, 1956.

Raman, C. V., "The heat capacity of diamond between 0 and 1000°K," *Proc. Indian Acad. Sci.* A46, 323, 1957.

Ramaswamy, C., "Infra-red spectrum of diamond by infra-red spectrometer and Raman methods," *Nature* 125, 704, 1930.

Ramdas, A. K., S. Rodriguez, M. Grimsditch, T. R. Anthony, and W. F. Banholzer, "Effect of isotopic constitution of diamond on its elastic constants: 13C diamond, the hardest known material," *Phys. Rev. Lett.* 71, 189, 1993.

Ruoff, A. L., and J. Wanagel, "High pressures on small areas," *Science* 198, 1037, 1977.

Sanchez, A., A. J. Strauss, R. L. Aggarwal, and R. E. Fahey, "Crystal growth, spectroscopy, and laser characteristics of Ti:Al$_2$O$_3$," *IEEE J. Quantum Electron.* 24, 995, 1988.

Sanchez, A., R. E. Fahey, A. J. Strauss, and R. L. Aggarwal, "Room-temperature continuous-wave operation of a Ti:Al$_2$O$_3$ laser," *Opt. Lett.* 11, 363, 1986.

Schmid, F., and D. C. Harris, "Effects of crystal orientation and temperature on the strength of sapphire," *J. Am. Ceram. Soc.* 81, 885, 1998.

Sobolev, N. V., Jr., and J. G. Lavrent'Ev, "Isomorphic sodium admixture in garnets formed at high pressures," *Contr. Mineral. Petrol.* 31, 1, 1971.

Sokhor, M. I., and V. D. Vitol, "X-ray investigation of thermal expansion in synthetic and natural diamonds," *Sov. Phys.–Crystallogr.* 14, 632, 1970.

Solin, S. A., and A. K. Ramdas, "Raman spectrum of diamond," *Phys. Rev. B* 1, 1687, 1970.

SPIE, "Ted Maiman and the birth of the laser," *SPIE Newsroom*, May 16, 2017.

Taylor, E. W., "Correlation of the Mohs's scale of hardness with the Vickers's hardness numbers," *Mineralogical Mag. J. Mineralogical Soc.* 28, 718, 1949.

Tefft, W. E., "Elastic constants of synthetic single crystal corundum," *J. Res. NBS (Phys. and Chem.)* 70A, 277, 1966.

Telling, R. H., C. J. Pickard, M. C. Payne, and J. E. Field, "Theoretical strength and cleavage of diamond," *Phys. Rev. Lett.* 84, 5160, 2000.

Thapa, J., B. Liu, S. D. Woodruff, B. D. Chorpening, and M. P. Buric, "Raman scattering in single-crystal sapphire at elevated temperatures," *Appl. Opt.* 56, 8598, 2017.

Thewlis, J., and A. R. Davey, "Thermal expansion of diamond," *Philos. Mag.* 1, 409, 1956.

Titkov, S. V., J. E. Shigley, C. M. Breeding, R. M. Mineeva, N. G. Zudin, A. M. Sergeev, "Notes & new techniques: Natural-color purple diamonds from Siberia," *Gems Gemol.* 44, 56, 2008.

Touloukian, Y. S., R. W. Powell, C. Y. Ho, and P. G. Klemens, "Thermophysical properties of matter – the TPRC data series. Thermal conductivity – nonmetallic solids," Office of Scientific and Technical Information (OSTI), USA, Vol. 2, 1971.

Turley, J., and G. Sines, "The anisotropy of Young's modulus, shear modulus and Poisson's ratio in cubic materials," *J. Phys. D.* 4, 264, 1971.

Victor, C., "Heat capacity of diamond at high temperatures," *J. Chem. Phys.* 36, 1903, 1962.

Wachtman, J. B., Jr., and D. G. Lam, Jr., "Young's modulus of various refractory materials as a function of temperature," *J. Am. Ceramic Soc.* 42, 254, 1959.

Wachtman, J. B., Jr., W. E. Tefft, D. G. Lam, Jr., and R. P. Stinchfield, "Elastic constants of synthetic single-crystal corundum at room temperature," *J. Am. Ceramic Soc.* 43, 334, 1960.